Titles of Related Interest

Cowboy Fiddler
by Frankie McWhorter
as told to John R. Erickson

Play It Lazy: The Bob Wills Fiddle Legacy
by Lanny Fiel and Frankie McWhorter,
with an audiocassette by Frankie McWhorter
produced by John R. Erickson

Horse Fixin'

HORSE FIXIN'

Forty Years of Working with Problem Horses

By Frankie McWhorter
As Told to John R. Erickson

TEXAS TECH UNIVERSITY PRESS

This book was set in 11 on 14 Bookman and printed on acid-free paper that meets the guidelines for permanence and durability of the Committee on Production Guidelines for Book Longevity of the Council on Library Resources. ∞

Art by Tom Floyd
Jacket and book design by Kelley Ferguson Farwell

Manufactured in the United States of America

Library of Congress Cataloging-in-Publication Data

McWhorter, Frankie.
Horse fixin' : forty years of working with problem horses / by Frankie McWhorter as told to John R. Erickson.
p. cm.
ISBN 0-89672-303-8
1. McWhorter, Frankie. 2. Horse trainers—Texas—Biography.
I. Erickson, John R. II. Title.
SF284.52.M38A3 1992
636.1'088'092—dc20
[B] 91-22595
CIP

Texas Tech University Press
Lubbock, Texas 79409-1037 USA

92 93 94 95 96 97 98 99 / 9 8 7 6 5 4 3 2 1

Introduction

This little book is the second of two books about the life and times of Frankie McWhorter—cowboy, fiddle player, and horsebreaker.

If you haven't read the other book, *Cowboy Fiddler*, you probably should, since it tells you more about the people and events that shaped the life of Frankie McWhorter than you will find in this one.

This volume is primarily a collection of his observations on horses: what they are, how they think, what makes them do the things they do, and how they can be trained to suit the needs of their masters.

Frankie came of age and learned his trade in an era that was very different from our own. As a teenager, he went to work as a cowboy on the sprawling JA ranch in the Texas Panhandle, when that ranch covered five hundred square miles of rugged country.

In *Cowboy Fiddler*, he notes that the JA horses "thought it was their first duty to kill a man." Oftentimes, when he got ready to mount his horse for the day, he needed the assistance of several strong men to hold the beast, so that he could climb aboard without being kicked and mauled. Once aboard, his first objective was to stay aboard and stay alive, and then to do his work. This is a sharp contrast to the kind of horsemanship we know today. Very few of us under the age of sixty have had that same experience, of riding out alone in a 120-section pasture on a beast that didn't like humans and wouldn't hesitate to hurt one if he got half a chance.

Frankie had that experience, and it has never left him. He knows the darkest corners of a horse's mind, and he knows firsthand what one can do to a fragile cowboy. The fact that he is still "fixing" bad horses in the sixth decade of his life suggests that he knows something that is worth hearing.

Another thing to remember about Frankie and the methods he uses is that *he seldom rides a finished horse*, the kind of mount you and I would ride *most* of the time. You and I ride a colt or a bronc only when we have to, and choose to do our work on an older horse that doesn't have to be schooled.

Over his lifetime, Frankie has reversed that. The teaching instinct is so strong in him that he is never out of school. The horses people bring to him, the horses he chooses to work with, are horses that other people have tried and failed to "fix."

That is what a horse fixer does. He takes the horses nobody else wants, and he tries to make honest animals out of them. If they can't be redeemed by a horse fixer like Frankie, they are useless and their next stop is the slaughterhouse.

If we were to draw an analogy to human behavior, we might say that Frankie has taken a special interest in delinquents and criminals. We wouldn't expect a prison guard to use the same techniques on his wards as a Sunday school teacher in the local Baptist church, and neither should we expect a horse fixer to approach his wards in the same way as a trainer who is dealing with docile animals that are eager to learn and please.

Communication is the objective in both cases, but what works for one might or might not work on the other. Frankie has found a way of communicating with the horses that come his way, and although his methods might seem rather blunt and harsh at times, they have worked.

Fifty years ago, his approach to horse training would not have seemed out of the ordinary. Back then, it was accepted that an undisciplined horse was worthless and that no horse had a right to ignore a reasonable command.

The difference between then and now is that today's horses are often objects of play rather than working tools. Many of the horses Frankie has fixed over the years have come from the race track, the show ring, and the rodeo arena. Most were born as good, honest horses and were taught bad habits by human masters who lacked either the will or the knowledge to educate them.

As Frankie notes, "Horses aren't born with bad habits. They learn them from people." Over his lifetime, Frankie has been very successful in dealing with problem horses. The modern reader, accustomed to dealing only with "nice" horses, might wince when he tells of tying a horse down and spanking him with a piece of chain, or shooting one with a BB gun.

Some might even accuse him of cruelty to animals.

I have known the man well, have seen him at work and at play, and I can assure you that cruelty is not in his nature. He is as big-hearted and gentle a man as I ever knew, and no one respects a horse or takes better care of one than Frankie.

But enough from me. This is Frankie's book, and it's time to listen to him talk about the subject he knows best and loves most: horses.

John R. Erickson

Contents

Horse Fixin'

Chapter One

Learning under Boyd Rogers

As long as I can remember, I've been associated with horses. My daddy was an excellent hand with a horse. He was a farmer but he knew how to turn one around. He knew how they thought.

One time he was whipping this mule with a piece of chain; I mean thrashing him. I started crying. I ran over and grabbed him by the leg to make him quit.

He told me, "Son, if you decide a mule or a stud horse needs a whipping, you tie him up and give him a *whipping*. And then leave him alone. Don't peck around on him and keep him mad all the time."

He never would let me ride a Shetland or a burro because they just kind of do what they want. He thought it would make me mean to my horses when I got older. He always put me on good horses, and I did the same with my boys.

One time Daddy was breaking horses down at Newlin, Texas, and this old man he was working for was a prankster. Name was Pistol Bill Rowell. Bill took me aside and said, "I'll give you a quarter if you'll throw a sack under that bronc." I did, and he bucked my dad off, and my dad did thrash me. Old Bill Rowell thought that was the funniest thing in the world.

Well, Daddy put me on a little bronc filly. Her name was Jenny Lynn. I won't forget that. He said, "Now son, I need to teach you something," and he throwed a sack under her legs and I got bucked off. And to this day, I've never bothered another man's horse and I sure don't like anybody messing with mine.

I started getting serious about horses when I was about fourteen years old. I was breaking horses down at Memphis, Texas, for those farmers, five dollars a head. One day in 1946 an old gentleman named Boyd Rogers saw me riding this little cold-blooded, short-necked colt, and I had it doing good.

He was from down around Nocona and he'd come up to the Panhandle to break horses. And he said to me, "Son, you sure have a nice set of hands. You don't pull hard on a horse." That didn't mean anything to me. I had pulled hard and it hadn't done any good. Well, he hired me to break horses for him and that's what got me to going.

Mr. Rogers and his wife raised some class horses on their ranch north of Memphis. They weren't considered class horses in Hall County because everyone around there was Quarter Horse people. But I guarantee you, if you went to New England and mentioned Boyd Rogers, people knew who you were talking about.

He was firm with a horse and he was the same way training me. He was a disciplinarian. There were times when he'd eat me out and there were times when he'd use a story and let me relate to it.

That was a smart old man. He graduated from Texas A&M in 1919 and he read all the time when he wasn't working. He'd never tell me I'd done something right. He'd say, "That wasn't bad, but if you'd done this, it would have been better." He'd never leave you with the feeling that you done something to perfection. He'd make you want to do more.

I'm the most fortunate person in the world to have been associated with him. He compared people to horses. Bob Wills compared people to mules.

That old man knew *so much* about a horse! One of the first things he taught me was about a horse licking his mouth. He had me driving horses

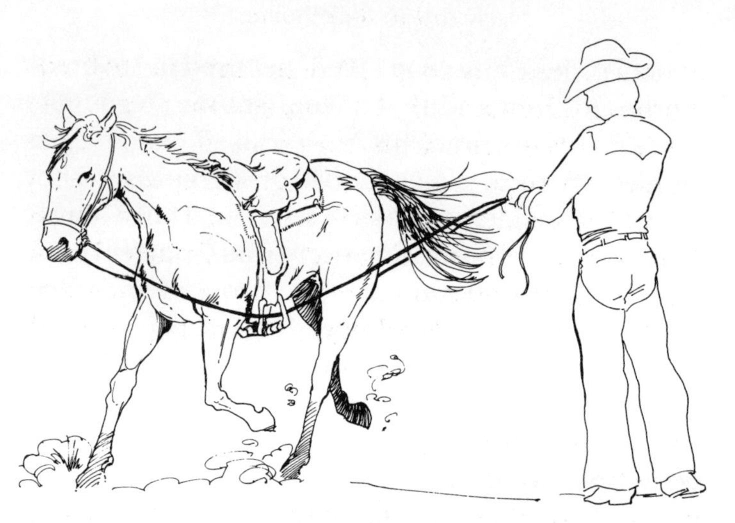

"I WANTED TO BE A WILD COWBOY AND I DIDN'T CARE MUCH ABOUT WORKING AFOOT."

with lines. I wanted to be a wild cowboy and I didn't care much about working afoot.

He'd watch what I was doing, and after a while he'd say, "Now stop him. Go up there and pet him, and stay there until he licks his mouth. If a horse ain't scared, he'll lick his mouth. If he is scared, you go ahead and pet him but don't spend a lot of time doing it because he won't lick his mouth until tomorrow or the next day."

He said, "They're just like scared people. The Indians respected a man who could spit around them. If he couldn't spit, they knew he was scared and he might wind up with his scalp hanging on a lodge pole. Horses are the same way."

Occasionally we'd get a horse that wouldn't lick his mouth at all, and Mr. Rogers would say, "He won't do. We don't need to spend a bunch of time on him. We'd just as well be playing dominoes."

After I went to work for him, he'd talk about that soft touch with a horse's mouth. He'd say, "You need to keep contact with a horse's mouth. Don't jerk on him. Pull and give him slack, pull and give slack. Handle him as though you had a pound of butter in your hands."

Another thing he told me was, "Don't ever dabble with your feet. When you kick one, he needs to grunt. If he doesn't, you might as well have not done anything. Let him know something's happening to him and then leave him alone."

Mr. Rogers used to talk about kids and broncs. He said, "There's no difference between raising kids and breaking broncs. Neither one of them knows anything until you teach them something, and what you teach 'em had better be right."

He had nothing good to say about children. They never had any children, and I mentioned it to him one time. He said, "The world is changing so fast and getting so corrupt, I'd hate to have it on my conscience for bringing a child into this environment they'll have to face later on."

But he just loved those little boys of mine and gave 'em both horses.

He kept books on every little thing. He kept a diary which was about half comical. "Tried to breed the Caspian Sea mare today but didn't." And, "Saw the first robin of the season and he's pretty cold."

He had a good sense of humor. He used to tell me that any intelligent person had a sense of humor. Or maybe what he said was, "Any person with a sense of humor is intelligent."

Mr. Rogers was good friends with Ben K. Green. Mr. Rogers had a high opinion of him, thought he was a fine person. And Mr. Rogers wasn't wrong too many times. And I'm sure Mr. Green thought the same thing about Mr. Rogers or he wouldn't have come up to Memphis to visit.

Ben used to come visit the ranch north of Memphis. I met him one evening. He'd just got there and I was just leaving. At the time, I was young and in a hurry and the name Ben Green didn't mean anything to me. I'd heard people talk about him, was all. Later on, when I was working on the JA ranch, Mr. Rogers brought him out to see me but I was gone.

Ben Green started the International Quarter Horse Association, and a lot of horses today could have passed his specifications, but back then there weren't many that could. Their necks were too short or they didn't have any withers or they were too wide between the legs. I believe there were twenty-two or thirty-two places where he measured them, and Doc Green did all the measuring himself.

I don't know why he couldn't get his association going. I guess he couldn't get enough members, but it was a step- ping stone for the American Quarter Horse Association.

Mr. Rogers raised straight Thoroughbred horses and made polo ponies out of them. That's how I got acquainted with the Thoroughbred's nature. When I was fifteen we went to New York with a load of polo ponies. I was tired of dragging that cotton sack around; I was ready to go do anything.

He had an old 1933 Dodge truck. We could only haul six horses in it. He told me I'd need a bedroll. I told Mother and she got me a tarp and a thin mattress and quilts. She being a mother, she gave me quite a few more quilts than I needed at that time of the year.

Mr. Rogers had taken the passenger seat out of the truck so he could haul saddles, and I rode from Memphis, Texas, to Ellicott City, Maryland, on a five-gallon bucket covered with a Navajo saddle blanket. We turned the truck over in Ellicott City, killed one horse, stayed in jail all night 'cause he didn't have any driver's license. I couldn't sleep. I'd never been in jail before. I was scared to death, eyes as big as saucers.

On the way up there, we stopped in Snyder, Oklahoma, and picked up a spotted stud, a little old cold-blooded–looking thing, no conformation, no personality. I had no idea why we were getting this horse. Mr. Rogers didn't say anything and I didn't ask him.

As soon as we got to New York, we started jumping him. The little old stud went to cheating him, like they do. So he decided he'd put me on him and pole him. That's where they tie a rope on the pole and when the horse goes over the jump, they jerk it up and hit him on the front legs with it and make him jump higher next time.

We got him jumping about 5 feet 8 inches and, heck, I couldn't ride him. I'd fall off when he'd hit the ground on the other side. He'd get right up to it and you'd think he was going to run through it, and then he'd just explode straight up.

Anyway, Mr. Rogers worked with him and got another guy to ride him, and he won the Madison Square Garden open jumping. He jumped 7 feet 4 inches. There was a wealthy Jewish fellow who owned the second-largest fur business in the world, and he'd been winning that thing. He wanted to know where that horse had come from and who owned him, and he and Mr. Rogers made a deal on him.

Another time we took sixteen head of horses back East in a railroad cattle car. We had built some stalls in there, but they put our car back by the caboose. You know how those old trains jerk around, and it's worse back at the end. By the time we got a hundred miles out of Memphis, all that jerking had tore those stalls down.

But we had this two-year-old colt, he was a race horse, and we had a blanket on him. Mr. Rogers would put me on him and I'd hold on to the front of the blanket while he led him, and we broke that colt to ride in the cattle car. He never bucked me off in the car, but going down the ramp when we unloaded them, say, he bucked me off several times.

But by the time we got up there with him, we had him broke to ride. He wasn't bridle wise, but he was used to having someone on his back. He was a tough little cookie.

The next year Mr. Rogers went down to South Texas and bought a bunch of them old ranch horses, the big ones (sixteen to seventeen hands) that they didn't like because they couldn't get on them easy. He was training them and me at the same time, and it was pretty hard on me. When we went over a jump, that little flat saddle would pinch their withers and they'd sure buck me off. He put a rope around their necks that I could hold on to, so I wouldn't jerk their mouths.

We jumped them through an alley, put them down this alley and he'd shoot them with a slingshot. We got them fixed and sold the fire out of them up there at Dedham, Massachusetts. I went fox hunting and I had a chance to stay up there and train horses for six hundred dollars a month. In 1947 that was pretty good wages for a sixteen-year-old boy.

But I didn't like the people up there, or very few of them. They made fun of me because I talked funny, so I came back home and went to work for the JA for ninety dollars a month, riding the craziest things in the world.

For Mr. Rogers, Sunday was just another work day. He didn't know Sunday from Wednesday. Mrs. Rogers went to church but we didn't. She'd want to take me to church but he'd tell her, "Sugar, I'd love to let him go, but we've just got too many things to do." And so that was out.

I broke a little mare for him one time. Her name was Dottie. I rode her for nine months or so, and she was a reining little darling. But she got in foal and he didn't get to take her with his polo pony string. He kept her and the next time he got her up to ride, she was in foal again. I don't know how that happened, but he finally just started breeding her.

One night, years later, I went to see him down at Nocona and he said, "I rode Dottie the other day, and it was just as though you'd gotten off of her yesterday. You really did a good job on her." That was his first and only compliment to me.

I asked where she was and he said, "She's right down here with the horses."

It was night and we pulled down there in the pickup. He called the horses and they came running in. Dottie ran through a board fence and ran a board right into her heart. Killed her as dead as a mackerel.

That old man sat down and cried like a baby.

As a young man, I got the reputation as someone who could do something with problem horses, and people would feed them to me. Since I was fifteen years old, people have been bringing me problem horses. They've always been a challenge to me, and still are. I can't do as much now as I did twenty years ago, or it takes me longer now. I've had to change my methods a little bit.

I would certainly not set myself up as an authority on horse training. All I know is what has worked for me. It seems that the more you learn about something, the more you realize you don't know.

Chapter Two

Training a Horse

The Proper Attitude

Horses are no different than people. If you take a man out and eat him up one side and down the other all day long, it's going to be drudgery for him.

It's the same way with a horse. If he starts to head a cow and you job a spur in him every time, before long he's thinking, "Well, why should I hurry? I'm going to get spurred anyway."

But if you brag on him a little, it makes a big difference. Maybe he doesn't go head that cow as fast as you think he should, but if you spur him about every third time instead of spurring him every time, I guarantee you, if he's the right kind of horse, the next time he'll head that cow quite a bit quicker.

Now, a horse can't reason. If he could, a man couldn't do anything with him. But they understand. They understand.

Early Training

The wrong training on the best-bred horse in the world will make him get worse instead of better. He could be twice as good as another horse, or twice as bad. The early training is very important. Don't ride him twenty miles and get him sore and dreading things. On a young horse, I try to make him look forward to seeing me. If you have to whip one around to make him go, you've done something wrong. He's confused. You've taken him too fast or asked too much of him.

Horse Training as Psychology

Training horses is all psychology. When they do something wrong and something bites them for it, like a spur or a bit, they need to think they did it to themselves. If they think *you* did it to them, it doesn't work as well.

If a horse won't turn to the left and you reach up and slap him in the eye, he knows you did it. But if you bite him somewhere, say with a slide bit on the corner of his mouth, and then wham that right spur in him, he can't see that you did it. All he knows is that he didn't turn and it hurt him in two different places.

You need to create the thought in his mind that it's his fault. Some of them take a little longer than others, but they will get it.

Bad Habits in Horses

Any bad habit in a horse is caused by people. They're not born with bad habits. A horse doesn't know anything until somebody teaches him something. A wild horse, like those old JA horses, they think they're protecting themselves from a man, just as they would from a mountain lion. It's a sixth sense with them.

When I break a young horse, I don't have any problems with him. When I get him broke, he doesn't have any bad habits. Or not many.

Breaking a Young Horse

There's more to breaking a young horse than just staying on his back. You've got to teach him something, and you can only teach him as much as he can absorb.

A young horse is like a child in kindergarten. What you teach him prepares him for the first grade. You *have* to teach a horse the basic fundamentals before you can go on to something else. You take things a step at a time, so that each step prepares him for the next one.

People would get mad if the teachers tried to make their kids do algebra in the second grade, but they sometimes expect that kind of progress out of a colt. You have to go through the steps. You don't teach one to stop out of a run. You teach him to walk and stop; then trot, walk, and stop; then lope, trot, walk, and stop.

It all falls into place and he'll stop from then on, or until someone ruins his mouth.

Once a horse accepts that you're the boss, he'll do more than he should because he wants to please you. If you let him do more than he should, the time will come when he goes backward in his training. When that happens, you almost have to go back to page one and start over.

Pushing a Young Horse Too Hard

The worst thing that has ever happened to the horse business is that a lot of people have invested big money in horses, and the owners are thinking about getting an immediate return on their money. That's legitimate in some areas, but when they start running races on yearlings, that's asking too much.

In the first place, a young horse isn't developed. If he runs hard, he's going to tear something up. All that punishment breaks down their knees. I'm not in favor of running horses until they're three-year-olds.

Years ago, when I was working on the JA ranch, they didn't expect a horse to do much until he was nine years old. Up until that time, you just rode him. They didn't think a horse was fully developed and ready to be trained until he was nine. They thought that pushing a young horse too hard too soon would ruin his legs, and if a horse doesn't have good legs, he doesn't have anything. It doesn't matter how smart he is.

Conformation

The physical qualities you look for in a horse depend on what you want to do with him. For a using horse, I look first at his eyes. If his eyes are big and honest, he'll have a good disposition. Then I look at his head, his neck, and his shoulders. I like a long neck with a thin throat latch, good withers, and a long sloping shoulder.

I like a nice long pastern that will make a horse flexible. A horse with sloping shoulders has long pasterns, and usually they're flexible and easy riding. A horse with straight shoulders and straight pasterns will eat your lunch. He'll give you indigestion after a hundred yards.

Those qualities are important, and a man needs to know what he's looking at. Judging a horse by the way he looks tells you that he's the right *kind* of a horse. Whether or not he turns out to be a good horse depends on what happens over the next six months.

A Horse's Eyes

Boyd Rogers, the man who taught me so much about horses, would size up a horse by studying its head, where the eyes were set, the size of the eye, and the expression in it. He'd say, "Now, that little old horse, you oughtn't to have any trouble with him. Look at his eye. He's got a big old eye. He's just confused. Don't be mean to him. Be nice to him."

There's no difference in reading a horse's eyes and people's eyes. *What you see is what they are.* Your first feeling is usually correct, and your experience with the horse will usually confirm it.

You should *never* look a nervous horse right square in the eye. I might glance at him to see where his ears are, but I never look with both my eyes into his eye.

If there's a horse you're afraid of, don't ever look at him with both your eyes. If there's a weak place in you, he'll find it. If you're worried about getting on him, he'll know it. If you're worried about riding him after you've got on him, he'll know it.

They read our eyes just like we read theirs. It's very interesting, and sometimes it's very embarrassing too.

I don't know how to explain that you can read a horse's nature in his eyes, but I derned sure know how to see it. It's there, if you know what to look for.

A Horse's Vision

Mr. Rogers used to tell me that a horse's eye magnifies a man seven times. A horse looks at you and thinks you're seven times bigger than you really are. Otherwise you'd look like an ant to him and he'd step on you. That's exactly what he'd do. The Lord was sharp enough to do that for us.

And when a horse sees a white rock in the middle of a road, his vision magnifies the rock, and he might think he can't get over it. That's why horses get spooky. They also have a blind spot.

"THERE'S A CERTAIN PLACE THEY HAVE TO GET TO SEE."

That's why you see cutting horses getting down low. There's a certain place they have to get to see.

A Horse's Sense of Smell

Horses can read your smell. You put off an odor and they can smell it and they know it. I've bluffed my way through a lot of things because I wasn't smart enough to be afraid. But as I got older and gained more respect, I think I began to put out an odor. And I started getting bucked off by horses that shouldn't have bucked with me, or wouldn't have, two or three years prior.

I really wasn't aware of it myself. Well, let's face it, I'd never tell anyone I was afraid of a horse. It ain't in my chemistry to say that. But I might say that I've developed quite a bit more *respect* for these horses.

There's not much you can do to hide your smell from a horse. If you ever get to the point where you're afraid of horses, you'd better take a job in a filling station and stay away from them, because they sense it. Or you might try riding older horses.

Trusting a Horse

These horses I work with today aren't nearly as bad as the ones I rode as a young man, but even today I don't leave myself vulnerable to a horse. I always treat a horse just like he was Twilight, the worst bronc I ever worked around. That has kept me from getting bunged up a lot of times.

One time I got off an old horse to go through a gate. I'd ridden this horse a long time. He was an Appaloosa horse a lady had given me. But the grass burrs were pretty tall around there and I got a head of them on my chaps.

When I got on him, those grass burrs hit him in the shoulder and he whammed me good. Before

"WHEN I GET ON A HORSE..."

it was over, I'd collected some more grass burrs. But that's what I mean. You can't get careless with one.

Mounting a Horse

When I was sixteen or seventeen years old, I was breaking Thoroughbred horses for Boyd Rogers. He'd throw the saddle on one that wouldn't do anything and he'd teach me how to get on that horse without the saddle being cinched up. No cinch at all. I can still do it today, although you need a horse with some withers on him.

He said, "You're learning where to put that weight so you'll have some advantage." I'm sure it's been a big factor with me because I've always been able to get on one. I might not be able to *stay* on him, but I can get on him.

When I get on a horse, I always turn my stirrup and run the left hackamore rein through my left palm and down to my elbow. And I take hold of his mane with the left hand and the saddle horn with my right. That way I can pull his head around to the left if he tries to whirl away to the right.

I had a horse on the Smith ranch, down at Childress. You had to take several wraps around your arm to hold him. He'd jerk the reins out of your hand. You couldn't hold him.

Leads

A lot of people can't tell when a horse is on the right or left lead, and many of the training books I've read don't explain it very well. One way you can tell which lead a horse is in is that one knee will creep forward, and that's the lead he's in.

It's a natural feeling. I can feel which lead a horse is on and I can feel it when he's getting ready to change.

Tie-downs

A lot of people use tie-downs today, even out in the pasture. I don't care for them. A horse doesn't have any freedom with a tie-down. If he starts to fall, a lot of times you can just throw him some slack and he'll use that head to balance himself and he'll get up. But if you've got his head tied down, he's liable to fall.

Stopping a Horse

The position of a horse's spine is very important. When you get into the finer points of training a horse to stop, you notice the position of his spine and nose. If his nose is down, it softens his shoulders and he usually can't stop on his front feet.

If he's loping, you count his feet—one-two-three, one-two-three—and you know where his foot is all the time. And when his hind feet are in the air, that's when you stop him, not when they're on the ground. You catch a horse when his front feet are on the ground to teach him to stop.

Fixing a horse so he won't stop on his front feet is one of the hardest things to do if he has been doing it very long.

Frankie McWhorter

Bits

A bit is kind of like a banjo or a fiddle. What you get out of it depends on who's holding it.

A bit to a horse is just like a pair of boots to a cowboy. Every one has a different feeling about it. Each one bites him in a different place. An old smart horse that's already had everything used on him, I might ride him with a different bridle every day.

Every horse trainer can find a use for corrective bits. You may have a problem almost solved and the horse figures out how to adjust himself on the bit you're using, and that's when a corrective bit comes in handy. You put another bit on him and it feels different, and while he's thinking about the new one, you've done what you've set out to do.

My main lick on a problem horse is that slide bit. It doesn't hurt the bars of his mouth and it doesn't hurt his tongue. It just works on the corners of his mouth. I had one old horse and the right side of his mouth was an inch longer than the left side, because he wouldn't turn to the right. But that will grow back.

The slide bit works completely different from a curb. With a curb bit, you pitch him slack when he slows down. That's his reward. But you hang on to the slide bit because you're not killing the nerves in his mouth. He may go fifty yards before

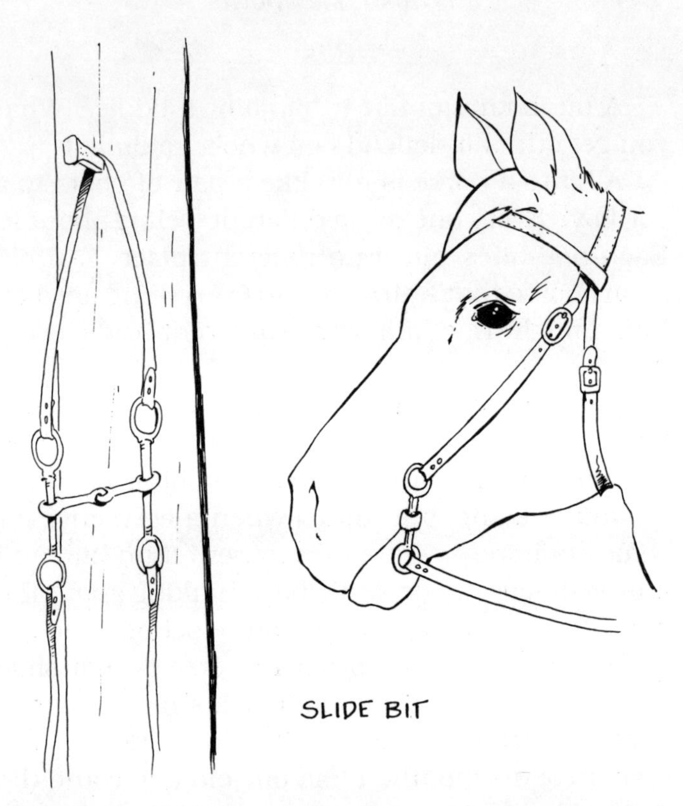

SLIDE BIT

he starts pawing at his ears, and when he does that, you're gaining. Tomorrow he'll do better, and after about three days, all at once you'll be loping along and he'll drop his rump into the ground. He thought you might have been ready to pull on those reins and he sure didn't want it to happen.

If I had to choose just one bit, I'd take the slide. It's as severe or as nice as you want it to be. Red

Snyder is the one who first showed me the slide and I went to a welder and had him make me up a pair. The shanks were made out of long nails.

The English Saddle

A lot of people might disagree with me, but I think you can tell more about what your horse is doing from an English saddle than from a stock saddle, because you're right down on him. They've gotten some of these cutting saddles cut down now where you can sure feel a horse.

You can learn to ride a horse in an English saddle. When you ride a flat saddle, you don't get careless or sloppy. When we were training jumping horses, Mr. Rogers would make me ride a flat saddle and go over a jump with my hands held out. He said, "You learn to ride a horse that way."

I rode a flat saddle for years when I was breaking Thoroughbreds. You can ride a hard-bucking horse in one of those things. As long as he bucks straight away, you can get a rein in each hand and pull them across your leg, and you can ride him. But if one ducks back, you're pretty fair game for him.

Mr. Rogers rode an English saddle most of the time. If he had a horse that he suspected would

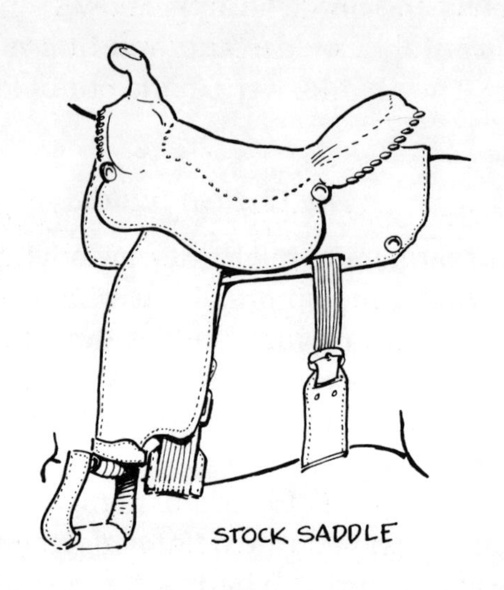

STOCK SADDLE

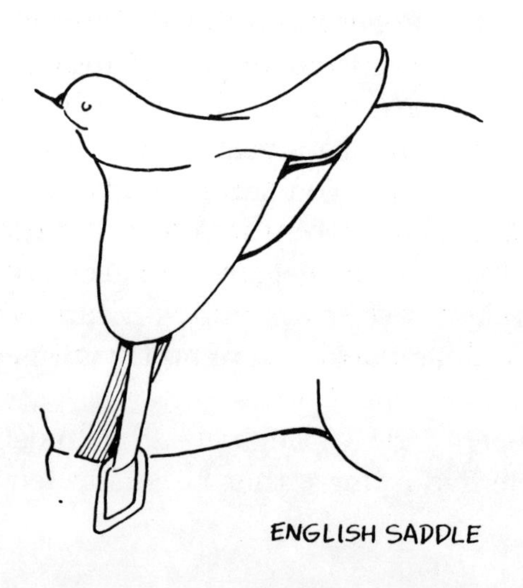

ENGLISH SADDLE

buck, he had an old McClellan saddle that he usec
He tied an old feed sack across the front of it.

Driving a Colt

Mr. Rogers made me drive a bronc with long reins. I didn't like it because people made fun of me: "What you going to do, ride him or plow the garden?" There I was, fifteen years old and wanting to be a wild cowboy—driving one of those broncs. As soon as I got away from him, I quit doing it.

But when I got older, I figured out that you can drive a horse ten saddles (ten times) and it's equivalent to riding him thirty days—if you don't get him mad or scare him or push him too hard. You hobble your stirrups (tie them together under his belly) and run the reins through the stirrups, get behind him and drive him. About the sixth or seventh day, you can pull his head around. Then you reinforce it with repetition, repetition.

By the time you get on him, he's already in the habit of doing it. When you pull on a rein, he turns without thinking about it. Very few horses that have been driven will buck with you.

"YOU CAN DRIVE A HORSE TEN SADDLES AND IT'S EQUIVALENT TO RIDING HIM THIRTY DAYS."

Frankie McWhorter

The Trainer's Age

I've noticed that my reflexes are getting slower with age. I can read a horse's mind as well as I ever could, but by the time the message gets to my hands to do something about it, I'm out of position.

No man who's gone through his life riding tough horses ever wants to think he's getting afraid of a horse, but it happens to everybody. Some of us won't admit it and some of us will. And when it comes, you'd better go to riding old horses that you're not uneasy about. You need to ride horses you've got a feeling about.

I'm a proud man and to say that I'm afraid of a horse would absolutely terrify me. People who know me don't associate me with fear. But the time will come. Age will see that it does because your reflexes can't handle the situation. Your mind is just as good. It ain't your mind, it's your reflexes.

I know what to do. That last old horse that bucked me off the other day, I knew what to do, but by the time I got my business in shape to stop it, he was too far gone. And so was I.

It's like going from being black-headed to gray-headed. Things change, and a person needs to accept that. No one wants to but you have to. That's the way it is.

Reading the Signs

If an old horse is going to buck you off day after tomorrow, he'll tell you *today* that it's going to happen. He'll do something.

I had a friend who used those seven-foot bridle reins. He had a little old filly there that was learning to foxtrot—a running walk. The little mare was give out and the boy had those bridle reins a-flopping around making her nervous. I said, "You'd better get some slack out of those reins, that thing's going to buck you off." He said, "That son of a gun can't find enough different positions to buck me off!"

A few days later, he called me again, wanted me to come over and help him. We rode up to a windmill to water his horse and I noticed that his bridle reins had been cut off. I said, "What happened to your bridle reins?" I figured he'd tied something up with them and had to cut them.

He said, "You know, this thing bucked me off the other day."

Teaching Aids

I never was a tough bronc rider. But I never really wanted to be a bronc rider. I wanted to learn how horses thought and then teach them something. Mr. Rogers kind of installed that in me. If you do the right things, it's easier for the horse to do what you want him to do than to do something else.

Just for instance, if you want him to turn to the right, you put your weight over that right hip pocket and he's going to have a little trouble trying to get that right hind foot out of the ground. It's easier for him to turn right than anything else.

Mr. Rogers told me one time, "If a horse is doing something that he knows better than to do, if you make it unpleasant enough for him, he'll try to see that it doesn't happen again."

I sure have enjoyed riding some pretty good horses because that old man told me that.

Appaloosa Horses

I'd never ridden any Appaloosa horses before I went to work for Tysons in Lipscomb County. I'd spent so much time making fun of them that I didn't realize they could be good horses. I don't know how many horses I broke over there. They raised a bunch of them and some of them went on and became good horses.

Those solid-color ones, we'd break 'em and use 'em. Those that were out of that old Double Six Domino stud all made cowhorses. They'd all watch a cow. But they'd also buck you off if you didn't kind of respect how they thought. I never had one buck me off, but they would if you just attacked them.

Thoroughbreds

I have found over the years that the only time Thoroughbred blood hurts a horse is when it ain't there.

You have to fix their minds early. They're different. They're excitable. The best horse I ever rode was a Thoroughbred named Danny Boy. I could rope mares off of him or put my kid on him, but still there was this little challenge going on between *us* all the time.

Like I told the man who bought him, "You don't *have* to take your spurs off, but you'd *better* if you don't want to get bucked off in front of a bunch of people." He would sure buck you off if you stuck a spur in him.

The reason cowboys have had trouble with Thoroughbred horses is that you can't punch one around or hit him. You can hit a Thoroughbred horse twice, but you'd better not hit him a third time. He'll do something about it. He'll fight you back one way or another, and usually he'll outsmart you.

Frankie McWhorter

Training Polo Ponies

Training polo ponies, I've loped a million circles. That's the way you fix a polo pony, repetition. Change their leads and fix their mouth. You can always tell about a polo pony. Those playing periods are seven and a half minutes long. If the game is exceptionally fast, after about six minutes of it you'll see him go to slapping his ears and looking to the barn.

That tells you his mouth wasn't completely made. He either hasn't had enough training or he's had the wrong kind of training.

Halter Breaking

I'll tell you what I got to doing when I worked over at the Tyson ranch. There was an old man over at Canadian named Ben Hill, raised a lot of horses. He'd halter break his colts by necking two of them together with a foot and a half of rope. I saw six or eight pair like that in a pen. They broke each other to lead that way.

I came home and tried it, but my rope was too long. One of them got a front foot over the rope. Now, it takes some doing to get that mess untangled!

A feller kind of needs to know the nature of the little whelps and put two of them about the same together. You don't want to put a real goosey, crazy one and a nice gentle one with a good eye together. You need to put two toughies together and two of those nice ones.

But it will work. It's kind of a breath-holding deal when you hook up that last snap. I've done it by myself. One man can do it alone, but he wants to be where he can climb a fence.

Ground Work

That horse's first impression of you is very important. He sizes people up just the way we do. If you hurt him, he'll associate you with hurt for a while. If he thinks you're in complete control of the situation, he'll respect you.

You shouldn't creep up on a horse like a mountain lion. Walk with some authority, like you're the boss, and go to his shoulder. If you appear to be in charge, he'll accept that. If you try to sneak up on a young horse, he won't let you because he's got an instinct to beware of predatory animals. An old horse will think you're afraid of him.

You go to his shoulder, not to his head, because a horse has a blind spot about four feet in front of him. If he loses sight of you in his blind spot, when he sees you again, it scares him.

I hardly ever hit a horse on the ground or do anything to make my presence unpleasant to him. If you whip them or run them around or rope them, they associate you with something unpleasant.

There's a place under a horse's belly where the flies get on him and he can't reach it with his tail or feet. If you find that place and scratch it, he'll be your friend.

Talking to a Horse

I like to talk to a horse. There's something about it that settles them, and they can read a man by his voice. That old Bernie horse of mine, I can say "Bernie" two or three different ways and he knows what I'm thinking.

Exercising a Horse's Neck

Old Boyd Rogers had fifty-five-gallon barrels for feed troughs and he'd bury them in the ground to that first ring. Then he'd pour some cement in the bottom and make those horses reach in there for their feed. It makes them use their necks and keeps the necks from getting big.

"IT MAKES THEM USE THEIR NECKS."

Chapter Three

Correcting Problems

Discipline and Firmness

There's always a point in a horse's life when you have to remind him that he's a horse and that he must do what he's told to do. I've never ridden a horse that didn't throw something at me, usually when I least expected it.

There's no way I could have a horse-training school to teach anybody because the Humane Society would be after me. Some of my methods are pretty crude, but they all boil down to one thing: *You're working on that horse's mind.* Even if you've got him tied down and you're whipping him with a rope, you're working on his mind, letting him know that where he is ain't where he wants to be.

But I love my horses, I do. They're just like people to me.

Barn-Spoiled Horses

One time I had a horse in Higgins, and his thing was going back to the house. Boy, he was a nice horse but he would go back to the house. One day he just about took one side out of James Robertson's barn.

There was an old kid there cleaning stalls. I tied the horse's front foot up and he lay down and I tied him down. I told this kid, "I'm going to get a cup of coffee and relax a little bit. I'll pay you ten dollars if every time you clean out a stall, you go out there and job that horse with a pitchfork and make him bleed." And he did.

Well, it was several days before that old horse would even *look* at that barn. Then he pulled it again and I did the same thing. That put a stop to it.

There are several ways of making the barn unpleasant for a horse. I have a paint horse called Flower. When I first started working with him, I'd put him in a stud pen and tie his old head just as high as I could. Then I'd go to town and eat dinner and leave him.

The worst thing you can do with a problem horse is to take him to the barn, wash him, and feed him. Sometimes I get down a quarter-mile from the barn, loosen him up, scratch him, and lead him on in. It works.

Old Bernie, my best using horse right now, when I first got him, you couldn't get him away from the house. He'd fall over backwards or buck or something. I put him in a stall, made a smooching sound with my lips, and shot him with a BB gun. For two or three days I'd saddle him and put him in there, and every time I went by the stall I'd smooch and shoot him.

I got on him in front of the barn one day and he went to rubbing his ears together and acting silly, and I made that sound. He ran 150 yards before he realized he was gone from the barn. I got off of him and petted him, scratched him, and talked to him. Then I led him back to that stall and shot him about ten times with that BB gun.

The next time, he went quite a bit farther. When he stopped and went to raising Cain, I got off, led him back, put him in the stall, and shot him again. Then I got me a little plastic sack and put some feed in it, and the next time out he went maybe a quarter of a mile. I got down and fed him some of that feed, and I've never had any trouble with him since.

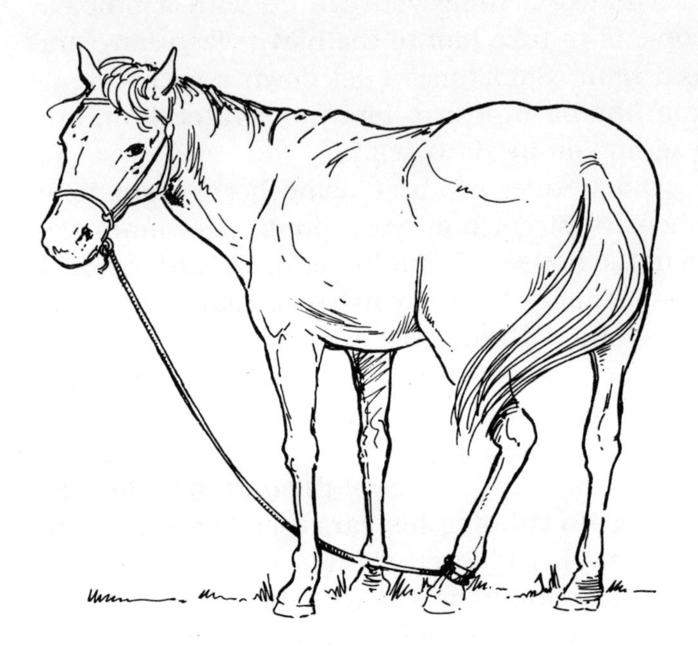

A HORSE THAT KNOWS BETTER.

Kicking

When I get a horse that's been handled and starts kicking at me—a horse that knows better, in other words—I put a rawhide noseband on him and tie a rope from the noseband to his hind foot. Then I leave him in the corral overnight.

Every time he kicks, he's jerking his head around. The next day, I can usually walk up and pat him on the behind without him kicking. The horse has punished himself for doing something wrong.

Biting

Some horses bite just because they're bored or because kids have poked them around with sticks. What I do to them is pull the whiskers on their chin every time I'm close to them, and sometimes I kick them on the front legs. What you're doing is distracting them before they decide to bite you.

If one bites me too hard or too often, I take him by the cheek of the bridle and kick him in the belly, until he respects me. That's why they do it, because they don't respect you.

A Nervous Horse

You should never pull on a nervous horse. That constant pulling is no good. You need to give him some slack.

See, slack is a horse's only reward for doing something right. If he isn't rewarded, in time he'll get to going farther and wider, and he'll learn to set his head where you can't control him. That's why I like to use that running martingale. With a tie-down (a standing martingale), a horse can learn how to get his head against that.

I never pull on one's mouth for long. I pull and then give him slack. It keeps his mouth alive. Instead of pulling one back all the time, you can make him go in a circle.

A Bucking Horse

There was never a prouder man than I was when I was 25 years old. I knew I couldn't ride a tough bucking horse, but I did know that I could ride a bunch of those horses that had bucked a lot of people off.

They wouldn't even buck with me. I knew how to keep them from it.

The timing is very important. You've got a split second to mess up his mind. He's thinking about bucking you off, and you know it, and you've got a short period of time to divert his attention to something else so that he doesn't really tie into it. You might pull his head around one way and then the other way very quickly.

When you get him diverted, then you do something else that messes his mind up even more. Once those suckers get their mind made up to buck, it's going to happen unless something occurs to change their train of thought.

I've said it before but I'll say it again. A horse can't think of but one thing at at time.

Head Slinging

If a horse is high headed, it's not his fault. It's some cowboy's fault.

A horse naturally uses his head to balance himself. When he stops, he may bring his head up but his nose isn't pointed straight up. What makes a high-headed horse is jerking on his mouth from a loose rein. They go to dodging.

What I do to fix a high-headed horse is use a slide bit that pinches the corners of his mouth, and run the reins through that running martingale, and I never give him any slack as long as his head's up there. He learns to drop his nose to get slack.

There was a horse on the JA's that I rode. The man who'd had him before me never gave a horse any slack, and that was the head-slingin'est son of a gun you ever saw. So one day I decided I'd fix him. I found a big steel nut and tied it into his foretop. The first time he tossed his head, that nut came down and hit him between the eyes. Liked to scared us both to death, but it sure stopped that head slinging.

You see, he thought he was doing it to himself.

"WHEN HE THREW HIS HEAD, THOSE FORKS OF THE STICK WOULD POKE HIM RIGHT BEHIND HIS EARS."

I worked with another horse when I was with Mendota Cattle Company on Red Deer Creek. Someone had tried to make a barrel horse out of him and they'd had his head tied down and he was breaking the tie-downs. I knew he'd make a good horse. He had a big old brown intelligent eye.

I was having to ride this horse to cover a lot of country, and I just hated to ride him. So one day I cut me a piece of cottonwood limb with a fork in it about five or six inches wide, trimmed it off, and held it so that when he threw his head, those forks of the stick would poke him right behind his ears. It sure worked.

A Horse That Won't Load in the Trailer

One day I was over in one of those south pastures on the Gray ranch. I'd taken Flower, that paint horse, and we'd ridden through some cattle and penned two yearlings in some pens, and I was going back over there and pick them up.

When we got back to the trailer, he would not load. He'd just go to backing up. I don't know what in the world went through his mind. I'd hauled him a thousand miles in that trailer and I'm pretty careful about throwing one around back there and getting him spoiled on a trailer. I've got

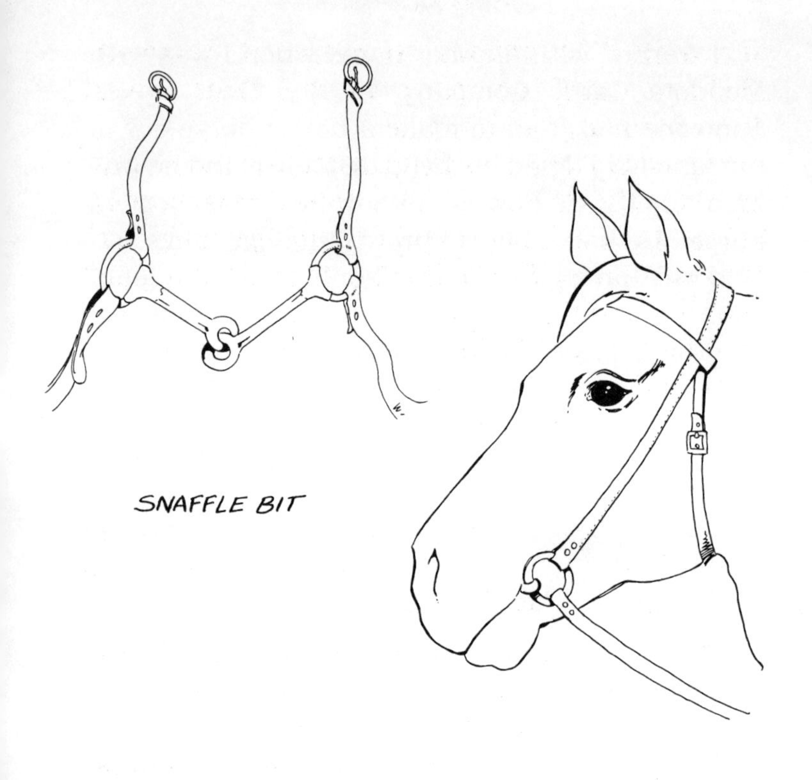

a lot of respect for a horse in a trailer. I've yet to this day to figure out why he did that.

I didn't have a halter with me. I most always carry a halter but I didn't have one this time. So I put my rope around his neck and shut the tail gate and tied him to it. Then I got my shovel out of the pickup and I did skin him up. I was mad, you know, and I'm sure I did more than was necessary. But anyway, when we got things settled down and I opened the trailer gate, he wanted in that trailer.

I loaded the steers in the front and put Flower in the back, and on the way home, every time I'd think about it, I'd stop and unload him and make him load again. I bet I did that twenty times going home.

Another horse, a brown horse, wouldn't load in a trailer, so I shut the tail gate and tied him to it, spanked on him and sicked my dog Hank on him. When I opened the tail gate again, he went right in.

But then I couldn't get him out, so I backed up to the corral fence. I was using a little snaffle gag bit that I use on a lot of young horses, with a nylon head stall and reins. I put a rein on each side of the saddle horn and tied my rope to the reins, got behind him and asked him to come out of there. When he didn't, I tied the other end of my rope to the bottom of one of the corral posts and drove out from under that little feller.

And pretty soon, after about three times, I could pull on those reins and tell him to get out, and he'd come out.

Habits and Horses

One of the first things Mr. Rogers told me was that everything a horse does is a habit. If you feed him in a certain stall three days in a row and then open all the stalls the next day, ninety percent of the time he'll go back to the stall he's used to. But if you don't build up the habit, he won't know where to go and he'll go into the first one he comes to.

It's the same thing if you're loping him in a right- hand circle and you let him lope with a left-hand lead in front. It'll get to be a habit and he'll keep on doing it.

Sometimes a horse will move around when you're trying to saddle him. If he moves around, I do something unpleasant to him and bring him right back into the same tracks he was in. He can only think of one thing at a time, and if you can divert his attention away from what he's thinking about, he'll start doing what you want him to do. And after he does that a couple of times, it becomes a habit with him.

Racehorses and Barrel Horses

Racehorses can be quite a problem, and so are horses that girls have tried to make into barrel horses. They're usually not horses with mean dispositions. You can handle them, they're gentle, but it takes longer to fix those two kinds of horses than any others I come across.

If six hundred horses went to the track in Raton this year, 580 of them weren't broke or even bridle-wise. The owners just want to get a run out of them and maybe don't care whether they're broke or not. And it's the same with barrel horses. They're run hard before they're broke.

I can fix those horses, but it takes a long time.

Changing Leads

In the last ten years I've fixed seven or eight racehorses that couldn't make the switch from a quarter-mile to a 550-yard track. The problem is that on the 550, he has to go around a turn and when he gets to that turn, he doesn't know how to handle it because he's never been around a curve.

Nobody had to tell me what was wrong. They were running on the wrong lead. The barn's always on the outside of that track and they're wanting to go to the barn. The horse will get on that righthand lead, wanting to go to the barn, and when he goes into that curve, he's got the wrong foot up. And he's got to slow down to change leads.

The position of their legs is SO important. All you've got to do is keep them in the proper lead and whap them good if they don't. A man who knows can see it coming. That ear will want to go over there, and when it does, that's when you do your thing. You have to keep him moving if he's on the proper lead.

I've fixed bunches of those. One of them I only had three days and the old boy took him up to Raton and won a big 550-yard race with him. He said, "I can fill your barn up with horses that need fixing." I told him my barn was as full as it needed to be.

Horses That Fall Over Backwards

Going over backwards is usually a horse's last ditch stand. If he's tried everything else and I've outsmarted him, he'll go over backwards.

Ninety percent of the time he does it when I'm trying to make him turn to the right, and most of the time he falls on his left side. If I can anticipate it and get him to do it, I can straighten him out.

You know when he's going to do it and you get ready for it. He'll tell you. If a horse's spine has a curve in it, even just one vertebra out of line, he can't rear up. His spine has got to be rigid in a

"HARD ON HORSES, EQUIPMENT, AND OLD MEN."

straight line. If you can get that nose over where you can see his eye, he can't rear up with you.

When I get one to go over backwards with me, I stand on the saddle horn until he quits struggling, if I can. Then I tie him down, go get a piece of chain, and spank him with it until he figures out that that is a very unpleasant place to be. Hopefully, he won't want to be there again.

Dealing with this problem is hard on horses, equipment, and old men.

Horses That Break Reins

It's hard to fix a horse who's developed the habit of breaking loose when he's tied. If he's sucessful breaking loose the first four or five times he's tied, it takes a long time to correct it.

Sometimes I'll put a war bridle on one and tie it to an inner tube and tie the inner tube to a post. I've also put a calving chain behind one's ears and tied the chain to a rope. I've also plaited a rope into one's tail, run it between his legs and through the halter, and tied it to a post. That tail-pu scares him when he goes backward. You can als put a loop around his belly and tie it, so that it wi draw down if he goes back.

Of them all, the calving chain has been most effective for me. It's kind of harsh and the ones I've used it on were left with a little white streak behind their ears, where that chain dug in. But you sure *could* tie them up after that with a string.

One time I got an old paint horse in, and they told me, "The horse is gentle, but whatever you do, don't tie him up solid." So the first thing I did was, I tied him to an iron post beside the barn. Instead of using a calf chain, I used a big windmill chain. He was a big old thing.

I laid my slicker there beside him, and for four days I left him tied and every time I went past him,

I'd just thrash him with that slicker. Lord, I thought he was going to tear everything down!

When we got done with that, you could tie him up with a cake string, but he had that white streak behind his ears. That chain dug into his flesh because he was serious about breaking away.

Breaking a-loose is a hard thing to fix, and most people won't go to the trouble. Of course, another way to solve the problem is never to tie one solid.

I don't like to tie up a horse unless I'm serious about it, and then I want to make sure I have the right equipment for it. Don't ever tie one up to or with anything he can tear up, and don't ever tie one down low. If they go to pulling up, they can pull something loose in their neck, and once they do that, their head's crooked and their lip falls. It ruins them. I never knew one to get over that.

Locoed Horses

I haven't been associated with a locoed horse in many years, but they're a whole new ball game. I used to run into it when I was young, down in Hall County. At first I didn't know what was wrong.

The first time it happened to me, I was riding this grulla filly. She was doing good. I could take her out there and lock her down. She'd stop and turn around. But then this loco thing came over her and she ran through a four-wire barbed-wire fence when I was riding her, cut her all to pieces.

The people who owned her went ahead and paid me my ten dollars for riding her, but they didn't let me ride her any more. They didn't think I was teaching her anything, but I was. I didn't know what was wrong, but later I told Boyd Rogers about it and he said, "She's had a bite of locoweed."

Loco is the first thing that comes out green in the spring, and that filly was running in a grazed-out trap. A hungry horse will get it because he's craving something green.

You might ride him for a week and everything will be lovely, and all at once he'll rub his ears together and fall over backwards for no reason.

It comes and goes. It seems to come when a horse gets hot. You might rope ten head of cattle on him and you're smiling because he's done so well, but when this thing hits him, he might run off a ninety-foot bluff with you. It's like getting off a good horse and getting on an idiot.

Before a horse has one of those seizures, you can see the signs that it's coming. When you're saddling him, he'll break out in big drops of sweat, as big as the end of a cigarette. They'll go to rolling off of him. If I ever see that again, I'll go to the house and play my fiddle.

I've always thought that loco impaired their vision. If one comes to a little cow trail, he might jump ten feet high to get across it. Or, more likely, he'll start running backwards and you cannot get him to cross it, no matter how much you whip him. You ride one up to a water trough and he'll go to drinking with his mouth ten feet before he gets there.

If you come to a gate with a piece of wire twisted between the tops of two tall gate posts, you can't get him to go through that gate. He might follow another horse through, but if you're by yourself, you might as well take off your hat and go to whipping on his face and back him through it. And even when you back one through, when he gets to where he can see that wire, you'd better have a hold of him because he's going to leave there.

I never knew of a horse getting over loco poisoning. You don't fix those horses. Normally, they fix *you.*

I haven't seen a locoed horse in fifteen years, and I think it's like coal oil lamps, a thing of the past. People today are more range conscious. They don't overgraze their pastures and eat them off, and horses are better fed now.

Snake Bite

A horse that's been snake bit and has gotten a good dose of that poison will behave like a locoed horse. When he gets hot, he goes to pot. He can't think, you can't control him.

I had one that I thought was locoed. He acted a lot the same. He'd dive off caprocks with me. But he went on to make a good horse, and I decided it was snake bite. He sure had a lot of respect for a snake. Since he made a good horse, I guess that means that snake bite isn't something permanent.

Frankie McWhorter

Horses That Are Hard to Catch

If you've got a horse that's hard to catch, you can spank his hind legs. Don't hurt him, just get his attention. Then back up, and pretty soon that horse will turn around and look at you. Never hit him when he's facing you and he won't fear the whip.

You can get the same results with a BB gun.

I made an almost-fatal mistake with a horse one time. I had this snakey little old thing. I knew he'd be a good horse. He had a good momma. But I couldn't catch him, I'd have to rope him to catch him, so I thought, well, I'll fix him. And I did, with a whip. Every time he'd turn his behind to me, I'd spank him across his back fetlocks and never hit him in front.

But what I didn't think about when I started driving him and he went to bucking, was that he thought he was supposed to come to me. I couldn't get away from him. It got pretty close. I finally got away from him, but that's the last one I ever tried to whip-break before I was done driving him.

Failures

Everybody has had horses he couldn't fix. I've had fewer than some people, I guess, but I've sent some home. I have *never* figured out how to fix a horse that fights a trailer.

It's near impossible to stop a horse from fighting a trailer, once he's got started. It does something to their mind. They don't want to be confined. Out here on the ranch, we don't have much of a problem with that because we use stock trailers, which are bigger than horse trailers, but you get one that's been slammed around in a small horse trailer and you've got problems.

Afterword

Readers of this book have commented that it ends abruptly. I know what they're talking about, and for a while it bothered me too. I decided to end it as is because *Horse Fixin'* is not a story; it's Frankie talking about horses, and when he's done, he's done. I couldn't think of a better way to end it and still can't.

John R. Erickson